Helpful Guide to Math
Solving Two Variable Equations

By

Craig Durfee

No part of this publication may be reproduced, stored in a retrieval system, or transmitted in any form or by any means, electronic, mechanical, photocopying, recording, or otherwise, without written permission of the publisher. For information regarding permission, write to mmtscraig@yahoo.com

Solving Two Variable Equations.

This guide will take you through a step by step process of solving two variable equations. The process will take you from solving a single variable equation to the two variable equation and explain each step along the way.

For teachers, this can be used as a good resource for lesson planning and for students you can follow along with Craig Durfee as he goes over it on youtube.

This is just the first book of many to help those in need of further instruction in math.

Enjoy and I hope this helps in your understanding.

Craig Durfee

Single Variable Equations

Solving Single Variable Equations

Before we start going over solving two variable equations, let us go ahead and revisit solving one variable equations using addition and subtraction.

x	+	5	=	8
x	+	5 -5	=	8 - 5
x	+	0	=	3
x			=	3

We are going to begin with this simple one variable equation.

$x + 5 = 8$

If we look at this you can probably solve this in your head but we are going to go through the normal steps to become familiar with solving equations.

The first step is to get x by itself. To do this we need to subtract five from both sides. It will look like this:

$x + 5 - 5 = 8 - 5$

$5 - 5 = 0$ and $8 - 5 = 3$

The new equation will then look like this:

$x + 0 = 3$

We will eliminate the zero because it has no value for this equation. The final equation is:

$x = 3$

> *How can we use this in a real-life situation?*
>
> *Imagine your school is selling a dozen cookies for eight dollars. You have five dollars in your wallet but that is not enough money. Your friend tells you he will give you the rest of the money if you will share some of the cookies. Now you are not going to write it out on paper but you will do the equation in your head. You know eight minus five is three. You tell your friend you need three dollars.*

Solving Single Variable Equations

Now let us look at a single variable equation that uses division and/or multiplication.

The equation for this example is:

$3x + 5 = 17$

$3x$	+	5	=	17
$3x$	+	5 - 5	=	17 - 5
$3x$	+	0	=	12
$3x$			=	12
$3x \div 3$			=	$12 \div 3$
x			=	4

We will look at this equation the same way we would looked at the previous one. The first step is to get x by itself. We need to subtract five from both sides.

$3x + 5 - 5 = 17 - 5$

After doing this it will then look like this:

$3x + 0 = 12$

We will eliminate the zero because it has no value for this equation. The resulting equation will look like this:

$3x = 12$

We will solve for x using division. We will divide by three on both sides.

$3x \div 3 = 12 \div 3$

We know $3 \div 3 = 1$ and we know $12 \div 3 = 4$.

Dividing by three will result in the final equation:

$x = 4$

> *How can we use this in a real-life situation?*
>
> *Imagine you're having a party. The restaurant has told you can only have seventeen people in their party room. Four of your friends said they were coming alone. Three said they want to bring other people. You need to figure out what is the maximum number of people they can bring. Four plus you will equal five. The variable x is the number of people each group can have. When you solve this, you will tell your friends they can bring three plus themselves in which becomes four.*

SOLVING TWO VARIABLE EQUATIONS   CRAIG DURFEE

Solving Two Variable Equations using Substitution

Solving Two Variable Equations using Substitution

We are now ready to look at two variable equations but you need to remember when you make an attempt to solve two variable equations you need two different equations.

2x	+	3y	=	23
2x	+	3y - 3y	=	23 - 3y
2x	+	0	=	23 - 3y
2x			=	23 - 3y
2x ÷ 2	=	23 ÷ 2	-	3y ÷ 2
x	=	11.5	-	1.5y

The two equations we will be using are as follows:

$2x + 3y = 23$
$5x + 4y = 40$

Just like when you are solving a one variable equation you will need to solve for x as your first step with the first equation.
To do this we need to subtract 3y from both sides. It will look like this:

$2x + 3y - 3y = 23 - 3y$

After doing this it will then look like this:

$2x + 0 = 23 - 3y$

We will eliminate the zero because it has no value for this equation. The equation will now look like this:

$2x = 23 - 3y$

At this point we will solve for x using division. We will divide by two on both sides like this:

$2x ÷ 2 = (23 - 3y) ÷ 2$

We will use the distributive property and it will look like this:

$2x ÷ 2 = 23 ÷ 2 - 3y ÷ 2$

Dividing by two will result in the final equation:

$x = 11.5 - 1.5y$

Now how does this help? What you need to know is that the value of x in both equations will always be the same and the value of y in both equations will always be the same. The next step is to substitute the value of x into the second equation. This will turn the second equation into a one variable equation.

Solving Two Variable Equations using Substitution

We now have a value for x that can be used to substitute into the next equation. This is what we have so far.

$x = 11.5 - 1.5y$
$5x + 4y = 40$

5x	+	4y	=	40
5 × (11.5 - 1.5y)	+	4y	=	40
57.5 - 7.5y	+	4y	=	40
57.5	+	4y - 7.5y	=	40
57.5 - 57.5	+	-3.5y	=	40 - 57.5
0	+	-3.5y	=	-17.5
		-3.5y ÷ -3.5	=	-17.5 ÷ -3.5
		y	=	5

The first step will be to put the value of 11.5 - 1.5y in for x in the second equation.

$5 \times (11.5 - 1.5y) + 4y = 40$

We will once again use the distributive property multiplying 5 with 11.5 and 5 with 1.5y

$57.5 - 7.5y + 4y = 40$

Combine like terms.

$57.5 - 3.5y = 40$

Subtract 57.5 from both sides.

$57.5 - 57.5 - 3.5y = 40 - 57.5$

The resulting equation is:

$0 - 3.5y = -17.5$

We divide by -3.5 on both sides and we come up with the value of y.

$-3.5y \div -3.5 = -17.5 \div -3.5$

We do the math and y – 5

We are not done yet. We came up with the value of y but we do not have the value of x. We need to substitute the value of y into one of the equations. Once again we end up with a single variable equation but this time instead of y it will be x.

Solving Two Variable Equations using Substitution

We are almost done. At this point, all we need to do is substitute the value of y into one of the equations. It doesn't matter which one we use but we need to pick one of them.

2x	+	3y	=	23
2x	+	3 × 5	=	23
2x	+	15	=	23
2x	+	15 - 15	=	23 - 15
2x	+	0	=	8
2x ÷ 2			=	8 ÷ 2
x			=	4

We will use the first equation.

$2x + 3y = 23$

Since y = 5 we will replace y with 5 in the equation giving us:

$2x + 3 \times 5 = 23$

The first step is to get x by itself. To do this we need to subtract fifteen from both sides. It will look like this:

$2x + 15 - 15 = 23 - 15$

After doing this it will then look like this:

$2x + 0 = 8$

We will eliminate the zero because it has no value for this equation. The equation will now look like this:

$2x = 8$

At this point we will solve for x using division. We will divide by two on both sides like this:

$2x \div 2 = 8 \div 2$

Dividing by two will result in the final equation:

$x = 4$

We came up with y and we came up with x but are we done? No, because you need to make sure the value of x and the value of y will work in the second equation.

Solving Two Variable Equations using Substitution

We used the first equation to solve for x and y but it is possible that the values we have will not work for the second equation. Hopefully, if you did not perform any mathematical errors then everything would be correct but you always need to check.

5x	+	4y	=	40
5 × 4	+	4 × 5	=	40
20	+	20	=	40

The second equation is:

$5x + 4y = 40$

We know from solving the first equation we said y = 5 and x = 4.

Let's now substitute 5 and 4 into the second equation.

$5 \times 4 + 4 \times 5 = 40$

We know 5 × 4 = 20 and we also know 4 × 5 = 20 resulting in:

$20 + 20 = 40$

Of course, simple math 20 + 20 is indeed 40.

How can we use this in a real-life situation?

Imagine that your boss is having a large meeting and wants to know if we order sandwiches and desserts from his favorite restaurant how much will the sandwiches be and how much will the desserts be. You look at the log for petty cash and you see two lunches at his favorite restaurant with the final cost and the tax for each. You can easily subtract out the tax and come up with the cost of the lunches. You also have the orders from when they asked you to call it in. The first order was 2 sandwiches and 3 desserts with a cost of 23 dollars. The second order was 5 sandwiches and 4 desserts with a cost of 40 dollars. From this you can figure out how much the sandwiches were and how much the desserts were. It would be impossible if you had only one order.

Is this the only way to solve for x and y? It is one way and I'll explain yet another way.

Solving Two Variable Equations using Addition and Subtraction

Solving Two Variable Equations using Addition and Subtraction

We are going to be looking at two variable equations but you need to remember when you make an attempt to solve two variable equations you need two different equations.

2x	+	3y	=	13
2x	+	7y	=	25
2x - 2x	+	3y - 7y	=	13 - 25
0		-4y	=	-12
		-4y ÷ -4	=	-12 ÷ -4
		y	=	3

The two equations we will be using are as follows:

$2x + 3y = 13$
$2x + 7y = 25$

The way we are going to do this is much different then we have done so far. With this we are going to subtract down. The resulting equation will look like:

$2x - 2x + 3y - 7y = 13 - 25$

Look what I just wrote and look at the two equations. See how I subtracted the second equation from the first equation. The end result will be:

$0 - 4y = -12$

We will eliminate the zero because it has no value for this equation. The equation will now look like this:

$-4y = -12$

At this point we will solve for y using division. We will divide by negative four on both sides like this:

$-4y \div -4 = -12 \div -4$

Dividing by four will result in the final equation:

$y = 3$

We just solved for y in one step versus two steps.

SOLVING TWO VARIABLE EQUATIONS — CRAIG DURFEE

Solving Two Variable Equations using Addition and Subtraction

We are almost done. At this point all we need to do is substitute the value of y into one of the equations. It doesn't matter which one we use but we need to pick one of them.

2x	+	3y	=	13
2x	+	3 × 3	=	13
2x	+	9	=	13
2x	+	9 - 9	=	13 - 9
2x	+	0	=	4
2x ÷ 2			=	4 ÷ 2
x			=	2

We will use the first equation.

2x + 3y = 13

Since y = 3 we will replace y with 3 in the equation giving us:

2x + 3 × 3 = 13

The first step is to get x by itself. To do this we need to subtract nine from both sides. It will look like this:

2x + 9 - 9 = 13 - 9

After doing this it will then look like this:

2x + 0 = 4

We will eliminate the zero because it has no value for this equation. The equation will now look like this:

2x = 4

At this point we will solve for x using division. We will divide by two on both sides like this:

2x ÷ 2 = 4 ÷ 2

Dividing by two will result in the final equation:

x = 2

We came up with y and we came up with x but are we done yet? No, because you need to make sure the value of x and the value of y will work in the second equation.

Solving Two Variable Equations using Addition and Subtraction

We used the first equation to solve for x and y but it is possible that the values we have will not work for the second equation. Hopefully, if you did not perform any mathematical errors then everything would be correct but you always need to check.

2x	+	7y	=	25
2 × 2	+	7 × 3	=	25
4	+	21	=	25

The second equation is:

$2x + 7y = 25$

We know from solving the first equation we said y = 3 and x = 2.

Let's now substitute 3 and 2 into the second equation.

$2 \times 2 + 7 \times 3 = 25$

We know 2 × 2 = 4 and we also know 7 × 3 = 21 resulting in:

$4 + 21 = 25$

Of course, simple math 4 + 21 is indeed 25.

Can we do this for the first set of two variable equations? Yes, but you have to do one extra step.

The equations were:

$2x + 3y = 23$
$5x + 4y = 40$

If we multiply the top equation by 5 and the bottom equation by 2 then it will look like:

$10x + 15y = 115$
$10x + 8y = 80$

You subtract and get:

$7y = 35$

$y = 5$

SOLVING TWO VARIABLE EQUATIONS — CRAIG DURFEE

I hope this has helped you with solving two variable equations. Look for my other books on algebra and my youtube videos going over two variable equations.